Data Science: Experiment, Validate, Collaborate

Data Science: Experiment, Validate, Collaborate

Chopra

First Printing, 2024

Horace, my friend, our journey has been a long and unexpected one

Contents

1 Introduction 1
1.1 Objectives . . . .
1.2 Document structure . . . .

2 Literature Review 3
2.1 Methodologies for ML projects . . . .
2.2 Technologies . . . .

3 COLLAPSE 15
3.1 Concept . . . .
3.2 Development process . . . .
3.3 Architecture of the COLLAPSE platform . . . .
3.4 Conclusion . . . .

4 Interface, Functional prototype and COLLAPSE effect 22
4.1 Interface design . . . .
4.2 Implementation . . . .
4.3 COLLAPSE effect - US25 . . . .
4.4 Conclusion . . . .

5 Tests and Results 36
5.1 Experimental conditions . . . .
5.2 Results . . . .

6 Conclusions and Future Work 45
6.1 Future work . . . .

Chapter 1

Introduction

Data science projects differ from the typical software development project because of its focus on experimentation, from trying different datasets, features and pre-processing tasks, algorithms and hyper parameters. This comes from the fact that data science projects apply the scientific method. This means asking a question, developing an hypothesis, testing the hypothesis and analyzing the results. Another major difference has to do with the validation of these projects. The validation of a typical software development project is mainly binary since it either works according to the requirements or does not. It is not as easy to decide this when it comes to data science projects since someone can argue that a model is working while another person can argue the contrary, and both can be correct given their frame of reference.

There are also conceptual and technological differences between software engineering and data science. This can be seen in the methodologies and technologies used. For example, versioning is a practice that would be considered a purely technological difference since the concept is common to both but the technologies used to achieve this might differ. On a regular software development project, code versioning can be achieved with git. A data science project can benefit from not only code but also data and model versioning. While code versioning can be covered by git, data and model can be better covered by tools like DVC, presented in section 2.2.3. Another example is experiment tracking, and it can be considered a conceptual difference since it is not present in SE practices but is vital for data science, given its focus on experimentation.

Since most projects are developed by teams, there is the need to guarantee that every member understands the present state of the project and what work has been done to reach said state. This includes keeping data about the sources used, the performed experiences and the obtained results by several members, so the team can compare experiments.

Because of this, several development methodologies specific for data science projects appeared and evolved along with the needs of the market [15] [1] [24]. The problem still lies in the fact that almost twenty years since its definition, the standard methodology for data science projects, CRISP-DM, has not evolved, and so, it does not provide any practices regarding, for example, data versioning. Once the standard could not keep up, several methodologies based on CRISP-DM appeared [15] in an effort to make teams adopt practices that would lead to better project development. The adoption of these methodologies brings the need to adopt some good practices [2] to ensure a successful project development [23].

1.1 Objectives

The objective of this project is to develop a tool that can aid teams in developing a data science project by improving collaboration and providing team members with an easy-to-understand, collaborative overview of the project's current state without being attached to any development environment or a specific language.

Besides improving collaboration, the tool aims to help with some important practices for data science projects: experiment tracking and results reproducibility. Experiment tracking allows teams to maintain a record of all the experiments performed, helping them avoid redundant experiments and facilitating the analysis of the project's progress.

Results reproducibility is essential to data science teams since it is necessary to be able to run an experiment numerous times (by several team members) and obtain the same results, not only so peers can review it but also to be able to compare different experiments.

Another objective is to validate the developed platform by enabling it to be used by data science teams on real projects. After using the platform, teams will be asked to respond to a survey so a performance analysis can be done on the platform's usage.

1.2 Document structure

In addition to the introductory chapter, this book includes three more chapters. Chapter 2 includes the literature review on development methodologies for ML projects and frameworks and tools to assist the development. Chapter 3 consists of an analysis of chosen architecture for the platform. Chapter 4 includes an explanation for the interface design and describes the features implemented. Chapter 5 consists of an analysis of the conducted usability tests. Chapter 6 includes conclusions taken from the developed work, as well as an indication of what the future work should be.

Chapter 2

Literature Review

This chapter presents the state-of-the-art of the topics related to this book. Since this book aimed at developing a tool for data science projects, it was necessary to develop a deeper understanding of the DM process and some of the most used methodologies. One of the objectives of this tool was improving collaboration in data science teams, so a round-up of the most common tools and technologies used by teams was done.

2.1 Methodologies for ML projects

The growth in size and, therefore, in the complexity of data mining projects over the years made it clear for data science teams that there could be a better way to approach the development process. The standard development methodology, CRISP-DM [15], has not been updated in nearly twenty years, so it begins to feel outdated against the current fast-changing market. A fast-changing market leads to the necessity of being able to update business requirements on the fly, by adding, removing, or changing them [6]. Software Engineering faced a similar problem and addressed it by developing Agile approaches [11], so it makes sense that data science teams did the same. The industry started to either adapt old methodologies or create new workflows that would be more Agile inspired [23]. With DS teams adopting more SE inspired methodologies comes the need to adopt good SE practices or even create new ones to ensure a successful development process. The following methodologies were chosen not only because they are amongst the most used but also because analyzing them as a group provides a good overview of the evolution of the methodologies used in data science projects. Starting with KDD, this analysis will cover a span of over twenty years.

2.1.1 KDD

The Knowledge Discovery in Databases [10] [24] is the process of using DM techniques to extract knowledge from data. The KDD is often considered a process and not a methodology. However,

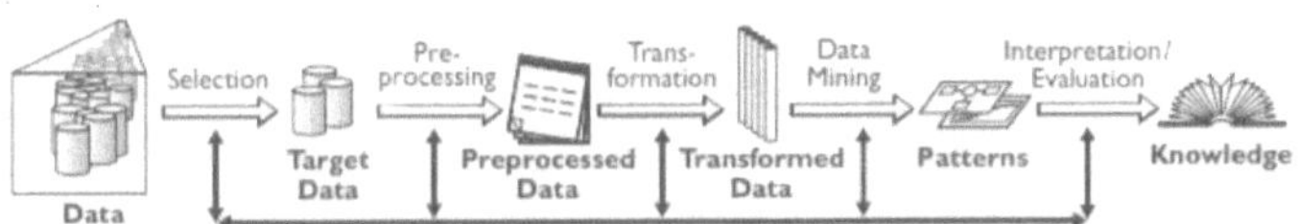

Figure 2.1: Simplified KDD stages. Source: [10]

the analysis of its stages is relevant to the topic since it can be considered the basis for several methodologies that have been introduced later. In this section, the analysis will be done to the KDD process as a nine-stage workflow and not the usual simplified five stages.

Developing and understanding of the application domain is the first stage of KDD, and it includes gathering relevant prior knowledge and understanding the application's goals.

Creating a dataset is the following stage and it is the selection of the dataset or focus on a subset of variables or data samples.

The **Data cleaning and pre-processing** stage includes removing noise and outliers, the decision of how to deal with missing data, and accounting for time sequence information. In this step, DBMS issues are also addressed, such as data types, schema, and mapping of missing values.

Data transformation aims to find useful features to represent the data based on the goal of the task and reducing the effective number of variables under consideration or finding invariant representations for the data using dimensionality reduction or transformation methods.

The **Choosing the suitable data mining task** stage is where the team decides the model's purpose like summarization, classification, regression, and clustering.

Choosing the suitable data mining algorithm is the next stage and it means choosing the appropriate algorithms and parameters to search for patterns in the data. It is also vital to match a particular data method with the overall criteria of the KDD process.

Employing data mining algorithm includes the search for patterns of interest in a representation or a set of representations such as classification rules or trees.

Interpreting mined patterns is the assessment and visualization of the extracted patterns, translation of useful patterns into user-friendly terms, removal of redundant or irrelevant patterns.

The last stage of one cycle of KDD is the **Using discovered knowledge** stage is where the team acts based on the knowledge extracted, integrating this knowledge into the performance system, or constructing a report to deliver to the interested parties. It is usual at this point to check for and resolve possible conflicts with existing knowledge.

2.1.2 SEMMA

The SEMMA process [20] [24] is a DM project development methodology developed by the SAS Institute. Even though the SEMMA process is independent of the DM tool chosen to implement it, it is often linked to the SAS Enterprise Miner. It comprises five stages: Sample, Explore, Modify, Model, and Assess.

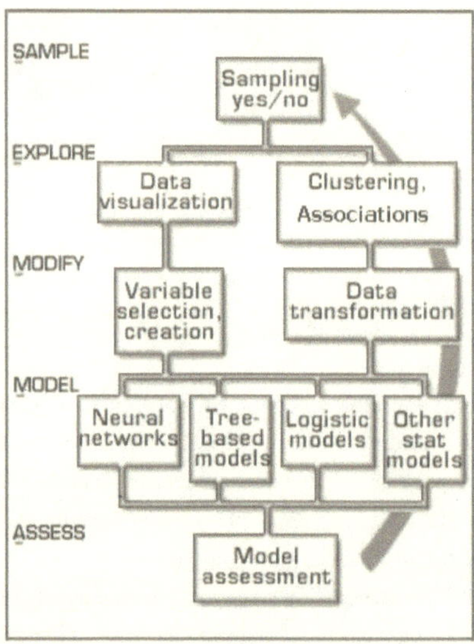

Figure 2.2: SEMMA process. Source: [20]

Often considered an optional stage, the **Sample** stage is where the sampling of data occurs. The objective is to extract a portion of a dataset small enough to be manipulated quickly yet large enough to contain relevant information. A pattern identifiable on the whole dataset is also identifiable on a partition that correctly represents the data. This is the stage where the data is split into training data, from which the models are generated, and validation data that is used later to test the performance of said models.

The **Explore** stage is the second stage in SEMMA. After the sampling stage, the objective is to explore the data, searching for trends, and irregularities to better understand the data. This data exploration can be done visually with visualization tools or numerically by analysing trends or groups.

The **Modify** stage involves transforming the data by creating, selecting, removing, or transforming variables.

The **Model** stage is responsible for generating the models. There are several modeling techniques available, each with its strengths and weaknesses, so the appropriate technique should be identified considering the problem in question.

The **Assess** stage involves the evaluation of the extracted knowledge on usefulness, reliability, and performance. This evaluation is usually done by applying the model to the test partition created in the sampling stage.

2.1.3 CRISP-DM

CRISP-DM or Cross Industry Standard Process for Data Mining is considered the standard methodology for the development of DM projects [29] [15]. The project lifecycle has six stages in CRISP-DM: Business Understanding, Data Understanding, Data preparation, Model Building, Evaluation, and Deployment.

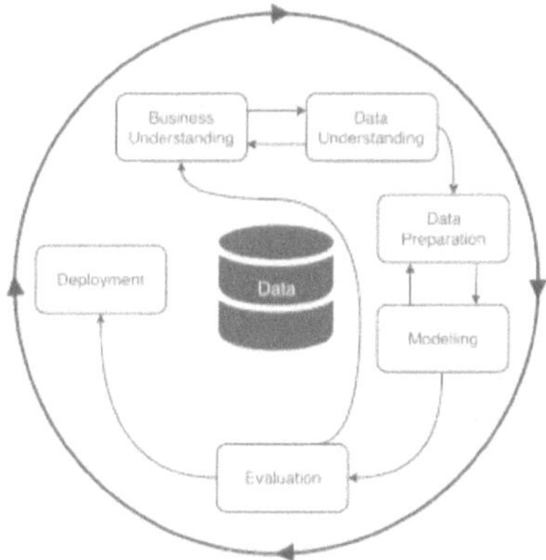

Figure 2.3: CRISP-DM lifecycle. Source: [15]

The **Business Understanding** stage's objective is to understand the project's goals and requirements from a business perspective. The next step in this stage is to use the knowledge extracted to define a DM problem and establish a development plan to meet the project's goals. Since the knowledge extracted form this stage can dictate a lot of business related decisions this stage is of extreme importance as it can determine the project's success.

The **Data Understanding** stage is responsible for handling the project's data-related requirements, from data gathering to data description and exploration and quality evaluation of the dataset. This is where the development team can apply tasks like clustering to better understand the data and possible underlying data relations.

In the **Data Preparation** stage, it is normal for the team to perform several tasks, including selecting tables, rows, attributes, feature engineering, and data cleaning to deal with missing values and outliers. This stage is, by nature, an iterative process. When working with data, it is normal for new problems to arise when performing a data preparation task, so the team usually performs each task several times to prepare the data for the modeling stage tools.

In the **Modeling** stage, for example, if the model has predictive purposes, it is imperative to split the data into training data, from which the models are generated, and test data that is used later to test the performance of said models. Several modeling techniques are selected and applied, having their parameters tuned to improve the obtained models.

The **Evaluation** stage is where the team evaluates the obtained models taking into account the business objectives defined in the first stage and while also assessing if any business requirements were not considered. The team can identify new patterns in the data, leading to new business requirements, making the development process go back to previous stages.

Considered the final stage of the process, the **Deployment** stage is where the team must report the knowledge extracted to the client in a user-friendly form as well as integrate the model in business processes. However, this stage should not be called final since alterations can occur that might call for the models' rework. For this, the models should continue to be monitored after the integration.

2.1.4 TDSP

The Team Data Science Process is "an agile, iterative data science methodology to deliver predictive analytics solutions and intelligent applications efficiently"[7] introduced by Microsoft in 2017 that also aims to improve team collaboration and learning.

The TDSP has four key components: a data science lifecycle definition; a standardized project structure; infrastructure and resources recommended for data science projects; and tools and utilities recommended for project execution.

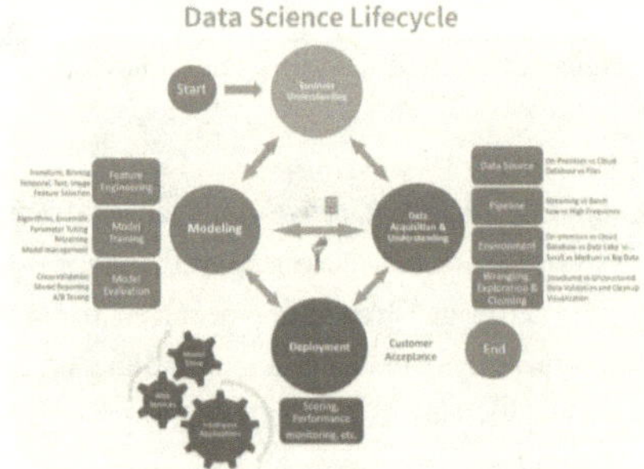

Figure 2.4: TDSP lifecycle. Source: [7]

2.1.4.1 Data Science Lifecycle

The **Data Science Lifecycle** identifies the major stages that projects usually execute in an iterative matter: Business Understanding, Data Acquisition and Understanding, Modeling, Deployment, and Customer Acceptance.

The *Business Understanding* stage aims to define objectives by working closely with the customer and stakeholders to comprehend and recognize the business problems. The team must form questions that represent the business goals that the data science techniques can target. It is likewise necessary to identify the data sources that the business has access to or needs to obtain. This stage has three artifacts associated with it: the *Project Charter*, considered a living document since it will be built upon during the development by updating it with new discoveries and changes to the business requirements, the *Data sources* document specifies the original and destination locations for the raw data and the *Data dictionaries* document that presents descriptions of the data rendered by the client.

The *Data acquisition and understanding* stage's goals are to construct a clean, high-quality dataset in which the relationship to the target variables can be easily understood and develop a solution architecture for a data pipeline that regularly refreshes and scores data. To accomplish this, the team must set up a process to gather the data from the source locations to the analysis environment, clean the data, audit its quality, perform data analysis to better understand the data's inherent patterns. After this, the team should determine if the data volume is enough to proceed to the modeling stage. This process is usually iterative, so there might be a necessity to discover new data sources with more relevant data to expand the dataset previously identified in earlier stages. The last step of this stage is to design the data pipeline. This data pipeline, developed in parallel with the next stage of the project, is usually either batch-based, streaming, or a hybrid, depending on the business need and constraints of the integration process. This stage has three artifacts associated with it: a Data quality report, a Solution architecture diagram or description for the pipeline, and a Checkpoint decision document containing where the team determines if they are ready to proceed, need to collect more data or abandon the project if the data is not sufficient.

The *Modeling* stage aims to generate an ML model that predicts the target with the best performance possible and suitable for production. To identify the optimal features, the team must apply feature engineering techniques. Next, the team should split the data randomly for modeling into training data and test data and then use the training data to build the models. The next step is to evaluate several competing ML algorithms with the various associated tuning parameters, also known as a parameter sweep. This stage's final step is to determine the "best" solution by comparing the algorithm's performance. This stage has three artifacts associated with it: Feature Sets report that contains information on how the features were generated; a Model report for each model generated with details on the experiment; and like the last stage, a Checkpoint decision.

The main goal of the *Deployment* stage is to operationalize the model. This means the deployment of the model and pipeline to a production or production-like environment. A way to deploy a model is to expose it with an open API interface to be easily consumed from various applications. This stage has three artifacts associated with it: a status dashboard displaying the overall system health and key metrics, a final modeling report with deployment details, and a final solution architecture document.

The last stage in the TDSP Data Science Lifecycle is the *Customer acceptance* stage, and its goal is to finalize the project deliverables. The team must validate the system, confirming that the deployed model and pipeline meet the customer's needs and performing the project hand-off. The only artifact associated with this stage is the Exit report containing all the information on how to run the system.

2.1.4.2 Standardized project structure

The second key component in TDSP is the **Standardized project structure**. TDSP recommends that all code and documents are stored in a Version Control System (VCS) like Git to facilitate team collaboration and implement task tracking. TDSP also provides a template for the project folder structure. These templates allow everyone to understand the work done by other team members and facilitate the addition of new members to the team.

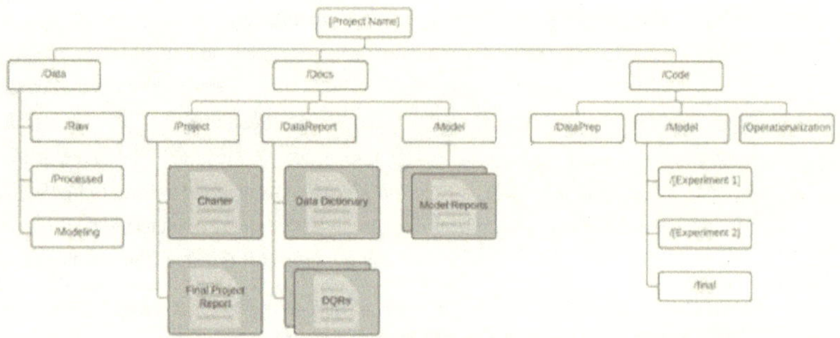

Figure 2.5: Recommended directory structure. Source: [7]

2.1.4.3 Infrastructure and resources for data science projects

The third key component of the TDSP is **Infrastructure and resources for data science projects**. TDSP recommends using cloud file systems to store datasets, databases, big data (SQL or Spark) clusters to manage shared analytics and storage infrastructure. An infrastructure like this enables reproducible analysis and avoids duplication, avoiding inconsistencies and unnecessary costs. Each team member should create a consistent computing environment so that various team members can replicate and validate the experiments.

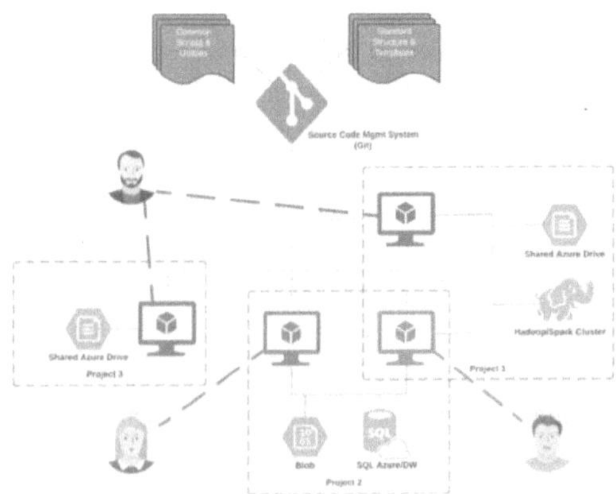

Figure 2.6: Example of infrastructure. Source: [7]

2.1.4.4 Tools and utilities for project execution

The last key component of the TDSP is **Tools and utilities for project execution**. Implementing a new process in organizations can be challenging. The use of the tools recommended by TDSP can help reduce the difficulty of these challenges and help automate some of the common tasks in the data science lifecycle like data exploration and baseline modeling.

2.1.5 Discussion

Table 2.1: Comparison of lifecycle stages

<table>
<tr><th>KDD</th><th>SEMMA</th><th>CRISP-DM</th><th>TDSP</th></tr>
<tr><td>Developing and understanding of the application domain</td><td>N/A</td><td>Business Understanding</td><td>Business Understanding</td></tr>
<tr><td>Creating a data set</td><td>Sample</td><td rowspan="2">Data Understanding</td><td rowspan="2">Data Acquisition and Understanding</td></tr>
<tr><td>Data cleaning and pre-processing</td><td>Explore</td></tr>
<tr><td>Data transformation</td><td>Modify</td><td>Data preparation</td><td rowspan="4">Modeling</td></tr>
<tr><td>Choosing the suitable data mining task</td><td rowspan="2">Model</td><td rowspan="2">Modeling</td></tr>
<tr><td>Employing data mining algorithm</td></tr>
<tr><td>Interpreting mined patterns</td><td>Assess</td><td>Evaluation</td></tr>
<tr><td rowspan="2">Using discovered knowledge</td><td>N/A</td><td rowspan="2">Deployment</td><td>Deployment</td></tr>
<tr><td>N/A</td><td>Customer acceptance</td></tr>
</table>

After an analysis of Table 2.1, it is possible to identify that the SEMMA methodology does not contain a stage dedicated to understanding the project's business requirements and a stage dedicated to the integration of the developed application into the business like the other methodologies. The five stages of SEMMA can be considered equivalent to the simplified five stages of KDD. An analysis of the comparison between stages can be done, using as reference the standard methodology CRISP-DM. The **Business Understanding** stage is equivalent to the TDSP stage with the same name and equivalent to the **Developing and understanding of the application domain** of KDD.

The **Data Understanding** stage is equivalent to the **Data Acquisition and Understanding** stage of the TDSP; it is equivalent to a combination of the **Sample** and **Explore** stages of the SEMMA and a combination of **Creating a data set** and **Data cleaning and pre-processing** stages of the KDD.

The **Data preparation** stage is one of the parts of the **Modeling** stage of the TDSP (along with the next two CRISP-DM stages), it is equivalent to the **Modify** stage of the SEMMA and equivalent to the **Data transformation** of the KDD.

The **Modeling** stage is equivalent to the **Model** stage of the SEMMA, and it is equivalent to a combination of the **Choosing the suitable data mining task** and **Employing data mining algorithm** stage of the KDD.

The **Evaluation** stage is equivalent to the **Assess** stage of the SEMMA and to the **Interpreting mined patterns** stage of the KDD.

The **Deployment** stage is equivalent to the combination of the **Deployment** and **Customer Acceptance** stages of the TDSP and to the **Using discovered knowledge** stage of the KDD.

Starting from KDD that provided a simple process to extract knowledge, to TDSP introducing a standardized structure for the project, it is possible to see the evolution into a more SE-like and Agile approach. Although at a high level these different methodologies have much in common, we can see an increase in the use of either old SE practices, adapted or new practices. Among these practices are practices like establishing and communicating clear objectives and metrics for training, versioning for data and models. It is possible to identify four significant effects of adopting these practices: Agility, Software Quality, Team Effectiveness, and Traceability[23].

2.2 Technologies

With the growing impact of DS projects came the need to have tools and frameworks that can facilitate the development of these projects. These tools and frameworks can improve agility, team effectiveness, the quality of the software produced, and team collaboration. This can happen by providing functionalities like task automation through pipelines, experiment reproducibility, and task tracking. The combined use of these tools and the previous chapter's methodologies can lead to a better development experience.

2.2.1 MLflow

MLflow [16] [5] is an open source platform for managing the end-to-end machine learning lifecycle and provides four primary functions: MLflow Tracking, MLflow Projects, MLflow Models, and MLflow Model Registry.

MLflow is not associated with any specific library so it can be used with any ML library and any programming language since the methods are accessible through a REST API and CLI. However, a Python, R, and Java API are also available. The **MLflow Tracking** component includes an API and a UI that allows the tracking of experiments to record and compare parameters and results by logging parameters, code versions, metrics, and output files.

An **MLflow Project** is a format for packaging data science code in a reusable and reproducible way, based on conventions. The Projects component includes an API and CLI tools for running projects, enabling the construction of workflows by chaining projects together. It is possible to run projects from a Git repository and deploy it on various environments like Databricks and Kubernetes.

An **MLflow Model** is a format for packaging ML models that can be, for example, real-time served through a REST API. This format defines a convention that allows a model to be saved in different "flavors".

The **MLflow Model Registry** component includes a model store, set of APIs, and UI to manage the full lifecycle of an MLflow Model collaboratively. It allows the team to know which MLflow experiment and run produced the model, model versioning, annotations, and stage transitions.

2.2.2 Kubeflow

The Kubeflow [14] [13] project aims to make the deployment of ML workflows simple, portable, and scalable by using Kubernetes. Kubeflow has several components from services for managing Jupyter notebooks, to the **Kubeflow Pipelines**, to other components like hyperparameter tuning and serving workloads across multiple platforms.

The **Metadata** component allows a better understanding and management of the ML workflows by tracking and managing metadata produced by these workflows. This metadata includes information about experiments or runs, models, datasets and files, and objects that compose the components' inputs and outputs in the ML workflow.

The **Fairing** component simplifies the building, training, and deployment of ML training tasks. It aims to package ML training tasks easily, enabling the packaging of the ML model training code and its dependencies as a Docker image. It also provides a high-level API for training models, making it easier to run training tasks in the cloud.

The **Feast** component is a feature store that allows users to define, manage, discover, serve, and validate features to ML models. A feature store is a system that helps to tackle some of the challenges that ML teams have when productionizing features like feature sharing and re-use, serving features at scale, consistency between training serving, point-in-time correctness, and data quality and validation.

The **Katib** component supports hyperparameter tuning for applications written in any language, supports early stopping and neural architecture search (NAS).

The **Pipelines** component is a platform for building and deploying ML pipelines based on containers. The Kubeflow Pipelines has three major goals: end-to-end orchestration, easy experimentation, and easy re-use. The first goal is achieved by enabling and streamlining the coordination of ML pipelines. The second goal, easy experimentation, is achieved by making it easy to try several approaches and techniques and manage these experiments. Kubeflow Pipelines achieves the third goal by enabling the re-use of components and pipelines to put together end-to-end solutions without rebuilding each time.

2.2.3 DVC

Data Version Control, or DVC [8], is a data and ML experiment management tool. DVC has four major functional components: Data versioning, Data access, Data pipelines, and Experiments.

The **Data versioning** component allows the team to use a regular Git workflow for large files, datasets, and ML models without actually storing large files in the repository.

The **Data access** allows the team to use data artifacts from outside the project and from another DVC project. This enables the download of a particular version of an ML model to a deployment environment and importing a model from another project.

Data pipelines provide an efficient way to reproduce models and data artifacts. DVC pipelines and their data can be versioned using Git, allowing for a better organization of projects, and results and workflow reproducibility. DVC data pipelines aim to solve some critical problems of ML projects like **automation** by automatically determining which parts of a project need to be run, **reproducibility** by storing in Git the files that describe which commands will generate the pipeline result. The DVC Data pipelines component also allows the visual representation of the pipeline in the terminal.

The **Experiments** component allows the team to easily try different approaches by tuning parameters, comparing the result's performance with metrics like AUC and precision, and obtain a visual representation of this with plots.

2.2.4 Discussion

All these tools provide features that can facilitate the development of DS projects as a team, not only by improving collaboration but by, for example, allowing results reproducibility, experiment tracking, or data versioning.

DVC is the only option that offers **Data versioning** functionalities. Moreover, its integration with Git makes it easy to deal with large files. It is possible to perform **Experiment tracking** in MLflow and Kubeflow. In MLflow, this is done by the component MLflow Tracking, and in Kubeflow, the Kubeflow Pipelines component provides a way to track experiments.

All the tools enable **Results Reproducibility** in some way. MLflow handles this problem by providing a way to package project code with the Projects component and its integration with the Tracking API component. The Kubeflow Pipelines component can be used to create workflows that are easy to reproduce. DVC handles this with the Data pipelines component.

DVC features enable teams to use versioning for several resources and make datasets available on a shared infrastructure. It provides a way to deal with large files, and its Git integration makes it so the data can be available for everyone in the company. Kubeflow lets teams log production predictions with the model's version and input data with the Metadata component. MLflow and Kubeflow both allow teams to share outcomes of experiments.

An opportunity to bring more value to the DS ecosystem presented itself when it became clear that despite these tools providing solutions for several problems related to the end-to-end life cycle of DS projects, they lack project management value. In addition, one of the issues identified with these tools is that it is not easy to have a clear overview of the project.

Chapter 3

COLLAPSE

This chapter describes the concept behind COLLAPSE, the development process of the COLLAPSE platform and the platform's architecture. This includes the specification of the methodology used, along with the phases of the development process. In addition, the architecture chosen for the platform is described, including the reasoning that lead to the decision of adopting the architectural style.

3.1 Concept

As seen in section 2.2.4, all the analyzed tools bring value to the DS ecosystem in their own way, helping users achieve data versioning, results reproducibility, experiment tracking and model deployment. However, these tools lack value when it comes to project management and providing all the information about the work done in a project in a single spot. A platform that tackled these issues would be meaningful and bring value to the realm of DS. Therefore, it was concluded that the place of the COLLAPSE platform in the DS environment is different from the tools previously analyzed. COLLAPSE will not try to do what these tools already do but will provide its value as a tool that will complement the already existing software presented in section 2.2 and support any of the methodologies presented in section 2.1.

The idea was to centralize all the valuable information about a project. This information needed to be organized, so it was divided into concepts. These concepts had to be generic and familiar enough for data scientists of all levels of experience, so, with this in mind, five concepts were defined: Data Sources, Data Preps, Datasets, Modelling, and Results. A Data source represents a raw source of data without any changes made by the user. A Data Prep (abbreviation for data preparation) represents any transformation the user applied to a Data Source in order to make it a Dataset. Therefore, a Dataset is what results from applying a Data Prep to a Data Source. A Modelling object comprises two parts: the code used to develop the model and the model itself. Finally, the Results resource, as the name suggests, represents the results obtained by a model.

It is noteworthy to mention here the work developed by Ana Sara Videira Morais in their book: "Data Mining Teams: Plataforma Colaborativa para Projetos de Data Mining" [17]. This book resulted in a collaborative platform for data mining with a heavy focus on the data aspect. Although COLLAPSE focuses more on providing an overview of the entire project, the influence of Data Mining Teams' (DMT) is noticeable when analyzing the relations between Data Preps and Datasets and Modelling and Results. The Data Preps and Modelling resources are tightly coupled with the Dataset and Results objects, respectively. This decision to focus on data in this way dictated the way several features were implemented. For example, only by editing a dataset can the user edit the data prep associated with it.

3.2 Development process

The COLLAPSE platform was developed with what can be considered a simplified version of the SCRUM methodology [22]. The SCRUM methodology is an agile software development methodology that breaks down the development into shorter development cycles, also known as *sprints*. In addition, agile methodologies focus on incrementally building the project, which allows the addition, update, or removal of features in the middle of the development. This enables a smoother and more flexible development process. The sprints occurred with a duration of one week to coincide with the weekly meeting with the supervisors. The development process started with the planning phase, where the most relevant features were identified and added to the backlog in the form of user stories. A user story is simply a planning tool, that allows the extraction of the features the software should have. Although the COLLAPSE platform is aimed at two types of users, Data scientists and Project managers, there is only one type of user in the platform to allow for more flexibility if one has to take the role of the other. This means that even though these users will focus on different features according to their role, they all have access to all the features. Since there is only one type of user on the platform, all the user stories defined on table 3.1 refer to a universal User.

Table 3.1: User stories

User story number	**User story**
US1	As a User, I want to create a new project, so that I can start the development process
US2	As a User, I want to view a project I am participating in, so that I can access its data
US3	As a User, I want to edit a project, so that I can update its data
US4	As a User, I want to create a new Data Source, so that I can share the resource with the team
US5	As a User, I want to view a Data Source, so that I can access the information about this resource
US6	As a User, I want to edit a Data Source, so that I can update its data
US7	As a User, I want to download a Data Source, so that I can analyze it in my device
US8	As a User, I want to create a new Data Prep, so that I can share the resource with the team
US9	As a User, I want to view a Data Prep, so that I can access the information about this resource
US10	As a User, I want to edit a Data Prep, so that I can update its data
US11	As a User, I want to download a Data Prep, so that I can analyze it in my device
US12	As a User, I want to create a new Dataset (with or without a Data Prep), so that I can share the resource with the team
US13	As a User, I want to view a Dataset, so that I can access the information about this resource
US14	As a User, I want to edit a Dataset, so that I can update its data
US15	As a User, I want to download a Dataset, so that I can analyze it in my device
US16	As a User, I want to create a new Modelling object, so that I can share the resource with the team
US17	As a User, I want to view a Modelling object, so that I can access the information about this resource
US18	As a User, I want to edit a Modelling object, so that I can update its data
US19	As a User, I want to download a Modelling object, so that I can analyze it in my device
US20	As a User, I want to create new Results (with or without a Modelling object), so that I can share the resource with the team
US21	As a User, I want to view a Results object, so that I can access the information about this resource
US22	As a User, I want to edit a Results object, so that I can update its data
US23	As a User, I want to download Results, so that I can analyze it in my device
US24	As a User, I want to add collaborators to my project, so that they can have access to the project
US25	As a User, I want to see all the resources associated with a specific resource, so that I can better understand the work already done by others

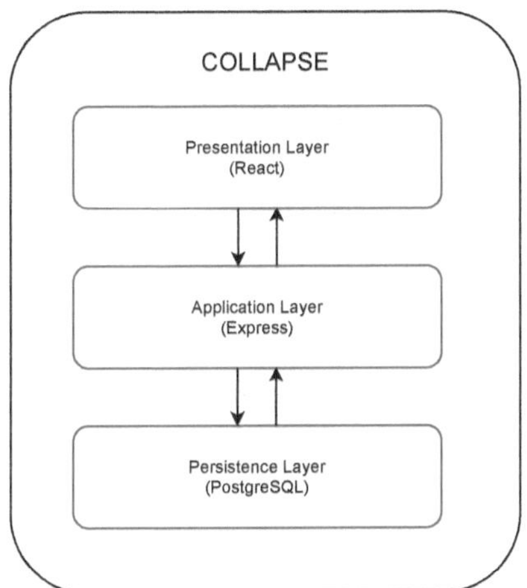

Figure 3.1: COLLAPSE architecture

The defined user stories were kept in mind when defining the architecture for the platform in case a specific architecture would simplify their implementation.

3.3 Architecture of the COLLAPSE platform

The architectural style adopted for developing the COLLAPSE platform was the layered architecture since it makes maintaining, testing, and updating layers separately relatively easy. To allow for a better understanding, the physical architecture and technological architectures will be presented at the same time. This means that a layer and the technology used to build it are presented at the same time.

The COLLAPSE platform can be divided into three layers: the presentation layer, the application layer, and the persistence layer. The presentation layer is responsible for providing an interactable interface to the user. The application layer establishes the connection between the presentation layer and the persistence layer. Finally, the persistence layer is responsible for connecting to the database, allowing the retrieval, insertion, update, and removal of the data inputted into COLLAPSE.

3.3.1 Presentation layer

Three technologies were compared to decide which one should be used for the presentation layer: React [18], Angular [3] and Vue [28]. These technologies were chosen because they are amongst the most popular, according to the StackOverflow 2020 Developer Survey [25].

The main deciding factor, given the time restrictions, was the previous experience with them. Due to the previous experience with these technologies consisting of only React, this technology was used as the basis of comparison. Besides the previously mentioned main deciding factor, React was chosen over Angular since it provides better performance [19], and has a bigger community [25], meaning that future developers that might continue this project will most likely be familiar with React, allowing for a smoother development. Therefore, the presentation layer was developed in React. Additionally, modularity and reusability were considered when designing and developing the components, allowing future developers to quickly understand and build upon the work already done. This layer also includes a thin sublayer in the form of a Web API. The COLLAPSE API was developed with a REST [12] architectural style. This style is widely used in web development, not only because it is easy to understand and implement but also because it makes the application more scalable.

3.3.2 Application layer

The application layer was built with Express [9]. Express is a free and open-source web application framework for Node.js used to build the backend of a web application, and was chosen since it is the most popular technology for this effect [25]. Express is simple to use, is supported by extensive documentation and has a large community, meaning that there are a lot of resources to help with debugging. Using Express allowed for smooth development since it only requires Javascript. In addition, this language was already being used in the presentation layer and is widely used in web development, so future developers can quickly learn it if not already present in their skillset.

3.3.3 Persistence layer

The persistence layer, sometimes called the storage or data access layer, is responsible for receiving data calls and providing access to the application's persistent storage. The concepts of COLLAPSE have, by nature, relations between them, so it was concluded that implementing a relational database over a non-relational database would be more appropriate. There are several database management systems. However, the decision was narrowed down to either MySQL or PostgreSQL, since they are the most popular [25]. Given the collaborative nature of COLLAPSE, it became clear that the chosen solution should be the one that handled concurrency the best since several team members could be trying to access the same resource simultaneously. Therefore, since PostgreSQL implements Multiversion Concurrency Control, it was chosen as the database management system for COLLAPSE.

The following tables constitute the database schema: Projects, Users, Data Sources, Data Preps, Datasets, Modelling, and Results. The database architecture is presented in figure 3.2.

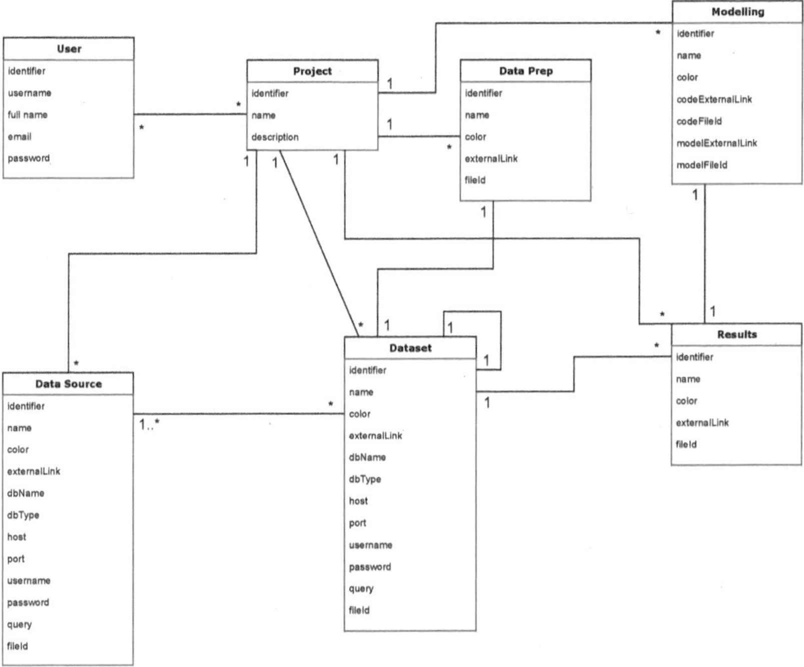

Figure 3.2: Architecture of the COLLAPSE database

3.4 Conclusion

This chapter focused on presenting the methodology used for the development process, the development process and its phases, and the platform's architecture. The chosen methodology was specified, along with the reasons why it was chosen. The planning phase was then described, identifying the user stories that the functional prototype should fulfill. This functional prototype is described in the next chapter. When it comes to the architecture of COLLAPSE, the layered architecture style was designated as the main architectural pattern. The three layers were defined, along with each layer's responsibilities. The technologies used for each layer were presented, along with the reasons why they were chosen.

Chapter 4

Interface, Functional prototype and COLLAPSE effect

This chapter presents the interface's design, evolution, the implemented features that constitute the functional prototype and the COLLAPSE effect.

4.1 Interface design

One of the main objectives of the COLLAPSE platform is to provide an easy-to-understand, almost "bird's-eye view" of the current state of the project. The platform was designed keeping this in mind since the beginning.

The mockups for the platform were created using Figma (figure 4.1).

The first sketch of this idea (figure 4.2) organizes these five concepts into columns. The five-column design ended up being used in the functional prototype. This type of design allows for all the valuable information to be centralized while being organized. This design is helpful for Project Managers and all the team members, allowing everyone to be aware of what has and has not been done, who did it, and how they did it.

4.2 Implementation

As mentioned before the user stories defined in table 3.1 served as a planning tool that helped identify the features that the COLLAPSE platform should have. In this section, each implemented feature will be matched with at least one user story.

Authentication - US26

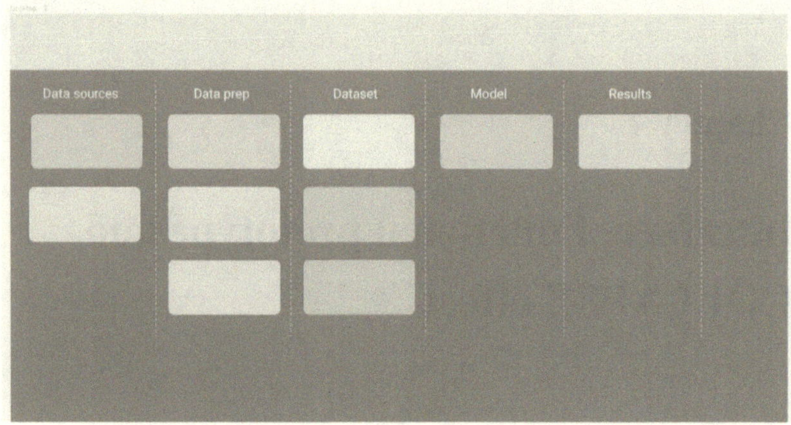

Figure 4.1: Mockups designed in Figma

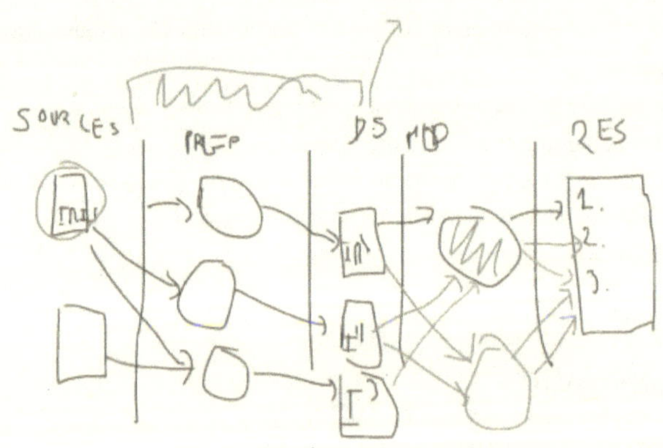

Figure 4.2: First sketch of the COLLAPSE board design

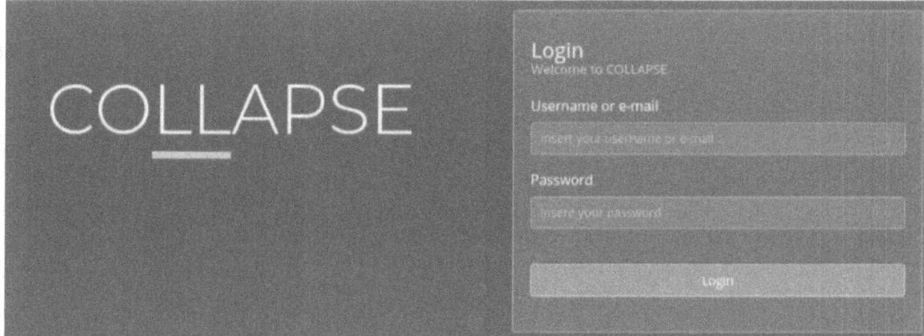

Figure 4.3: Login page

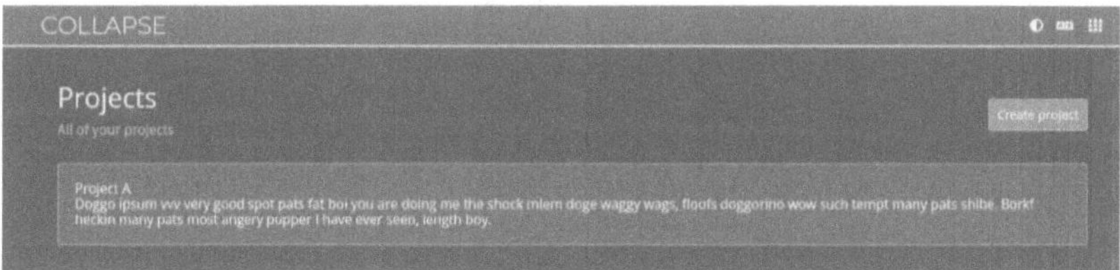

Figure 4.4: Projects page

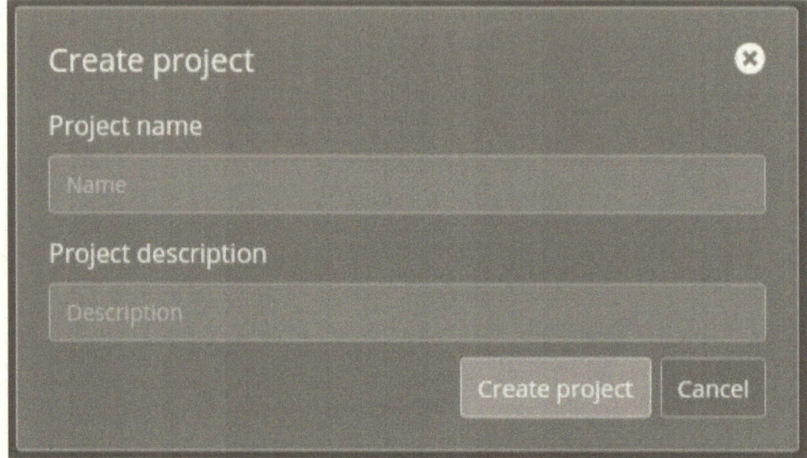

Figure 4.5: Projects creation modal

For a user to have access to the projects they are participating in, they must be authenticated. The authentication procedure is familiar to most users. Upon reaching the Login page (figure 4.3), the user has to supply their email address or username, and password. After clicking the login button, the user will be redirected to the Projects page (figure 4.4 - US2) to see all the projects they are currently working on. At any time, the user can click the icon in the navigation bar and logout (US 27).

Creating a project - US1

While on the Projects page, the user can create a new project by clicking the "New project" button (figure 4.6). This will open a modal asking for the project information (figure 4.5).

Navigating the board - US5, US9, US13, US17, US21

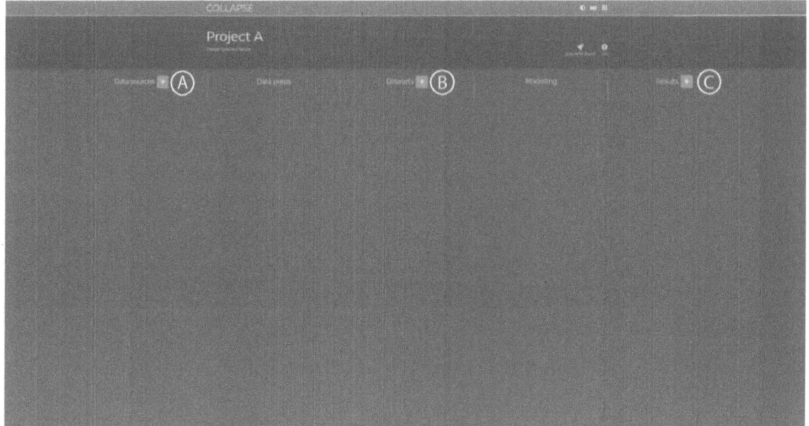

Figure 4.6: COLLAPSE board

After clicking a project card the user is redirected to the Project page, where they are presented with the COLLAPSE board (figure 4.6). All the members of the project share this board to improve collaboration.

Creating a data source - US4

Figure 4.7: Data Source creation modal

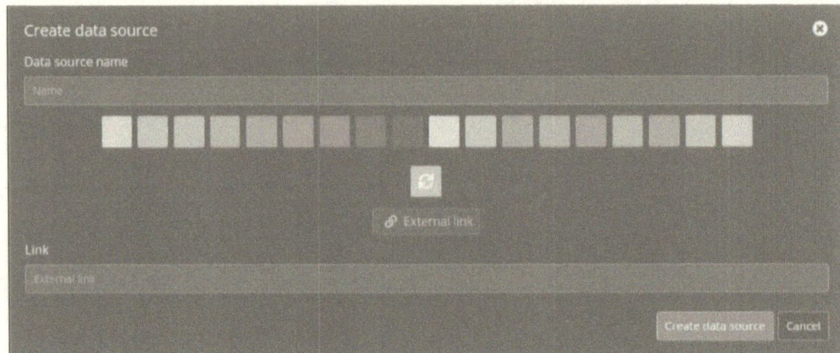

Figure 4.8: Creating a Data Source with a link

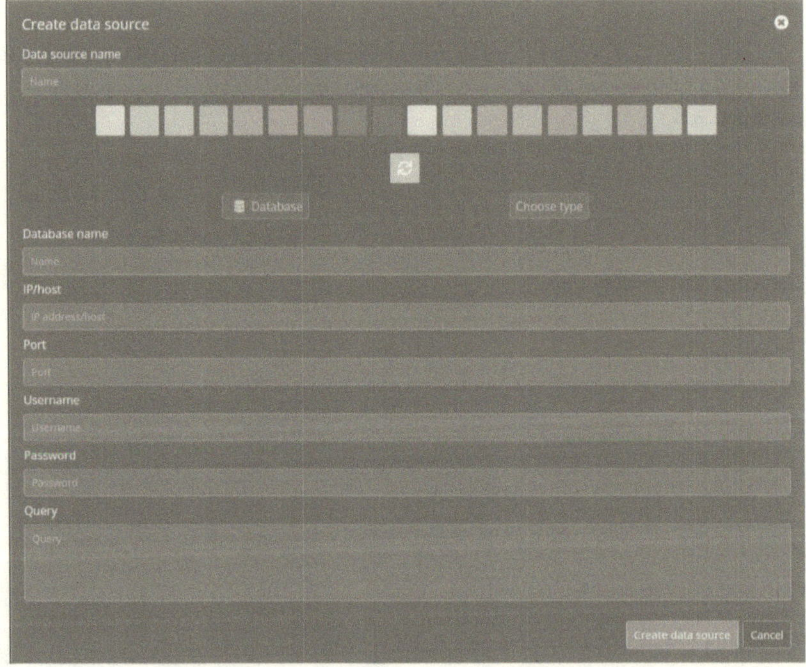

Figure 4.9: Creating a Data Source with a database connection

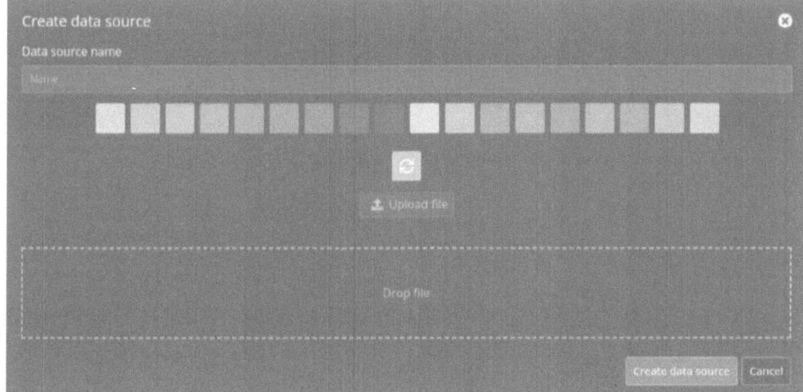

Figure 4.10: Creating a Data Source by uploading a file

After clicking the button to add a new data source (figure 4.6 A), the new Data Source form (figure 4.7) will be presented to the user, where they can provide some information and create a new data source. This form changes depending on the type of source the user chooses. The creation of a data source can be done with an external link (figure 4.8), a database connection (figure 4.9), or by uploading a file (figure 4.10).

Creating a dataset - US12, US8

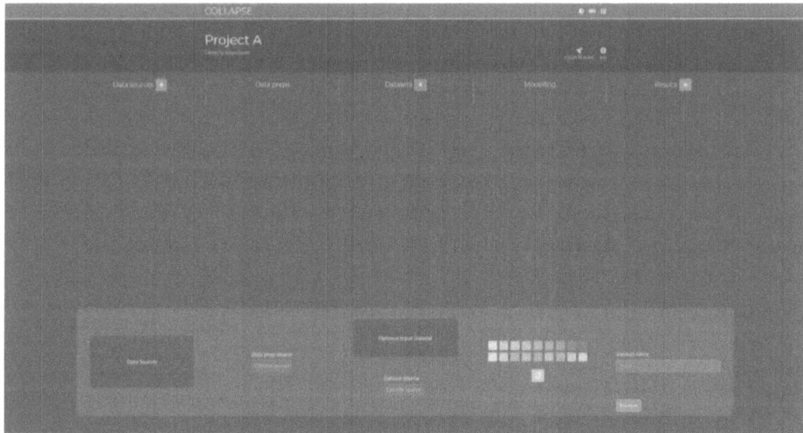

Figure 4.11: Dataset creation form

Figure 4.12: Adding a Data Source to the Dataset form

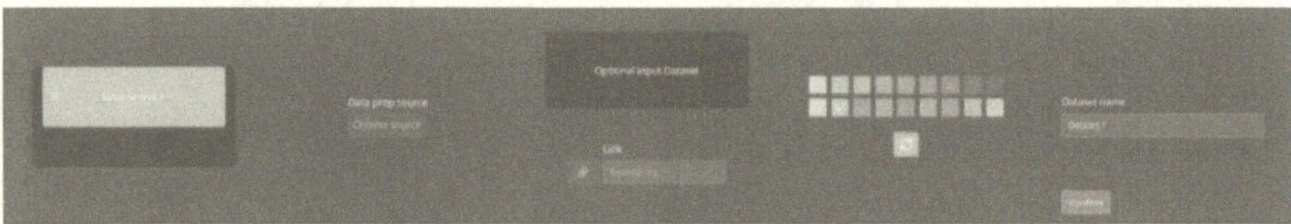

Figure 4.13: Creating a Dataset with a link

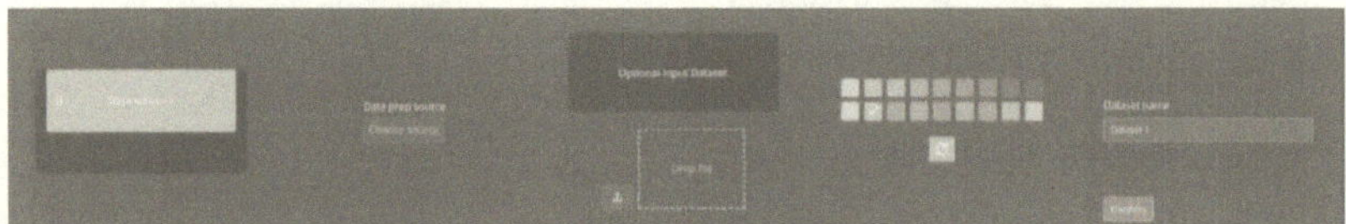

Figure 4.14: Creating a Dataset by uploading a file

After clicking the button to add a new Dataset (figure 4.6 B), the new dataset form (figure 4.11) is presented to the user. Next, the user must drag and drop a Data Source to the droppable area (figure 4.12) and provide a source for the Dataset itself. The Data Prep input is optional since the user may want to share the Dataset with team members but not share the data prep as it may not be ready yet. The Dataset resource can be created with a link (figure 4.13) or by uploading a file (figure 4.14).

Creating results - US20, US16

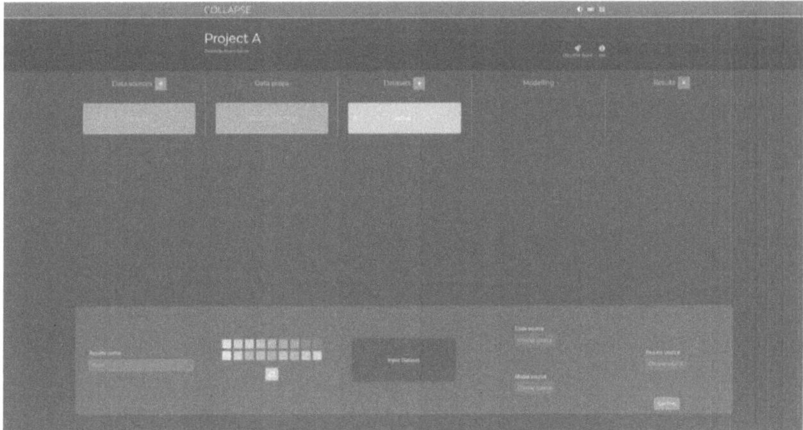

Figure 4.15: Results creation form

Figure 4.16: Adding a Dataset to the Results form

Figure 4.17: Creating Results with a link

Figure 4.18: Creating Results by uploading a file

After clicking the button to add new Results (figure 4.6 C), the new results form (figure 4.15) is presented to the user. The user must drag-and-drop the Dataset (figure 4.16) used to achieve the results and provide a source for the results. The source of the results can either be a link (figure 4.17) or an uploaded file (figure 4.18).

Editing a resource - US6, US10, US14, US18, US22

Figure 4.19: COLLAPSE card with Edit action

Figure 4.20: COLLAPSE card without Edit action

The user can edit the resource by clicking the editing action when hovering over the card (figure 4.19). This will open the same form used in creating the resource but with the fields already filled with the resource's information. Some resource cards like the Data Preps and Modelling will not have this action (figure 4.20) since they are tightly coupled with the Dataset and Results objects, respectively. To edit the source of a Data Prep, the user must edit the Dataset where the Data Prep in question was used. The same applies to when a user wants to edit the sources for the Modelling object, meaning that the user must click the editing action on the Results card to edit the Modelling sources.

COLLAPSE card actions - US7, US11, US15, US19, US23

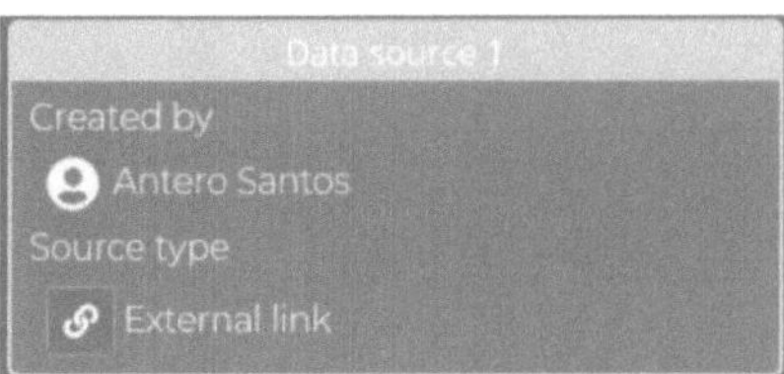

Figure 4.21: Expanded COLLAPSE card

The COLLAPSE card offers more actions than the editing action, also allowing the user to download the resource or examine it. The downloading action can have three different behaviors depending on the type of source the resource has. If the resource was created with an external link, a new browser tab will open on that link upon clicking the action. Suppose the resource was created with a database. In that case, the platform will attempt to use the values provided by the user to establish a connection to the database, execute the query and download the result of the query in the form of a .csv file. The remaining action is the examining action. Upon clicking the card, the card will expand (figure 4.21), revealing the responsible for its creation and which type of source it has.

4.3 COLLAPSE effect - US25

Although this was one of the features that could not be implemented it is still worthy to mention. The COLLAPSE effect is the feature extracted from US25 (table 3.1). The idea was that once the user clicked the COLLAPSE action on the card, every resource (Data Source, Data Prep, Dataset, Modelling or Results) associated with this resource would be highlighted, allowing the user to quickly visualize all the work related to it.

Taking for example, a COLLAPSE board in the state of figure 4.22. The user wants to view all the resources associated with the dataset on the second row (orange), so the user hovers over the card, clicks the action and the COLLAPSE effect occurs, resulting in the state represented in figure 4.23 All the resources associated with the dataset are brought to the top of the board and the remaining resources are greyed out. The concept of the associated resources varies from resource to resource. In this example, the dataset, these resources could be the data sources used to create this dataset, the data prep linked to it, the results created with the dataset and the modelling objects linked with the results.

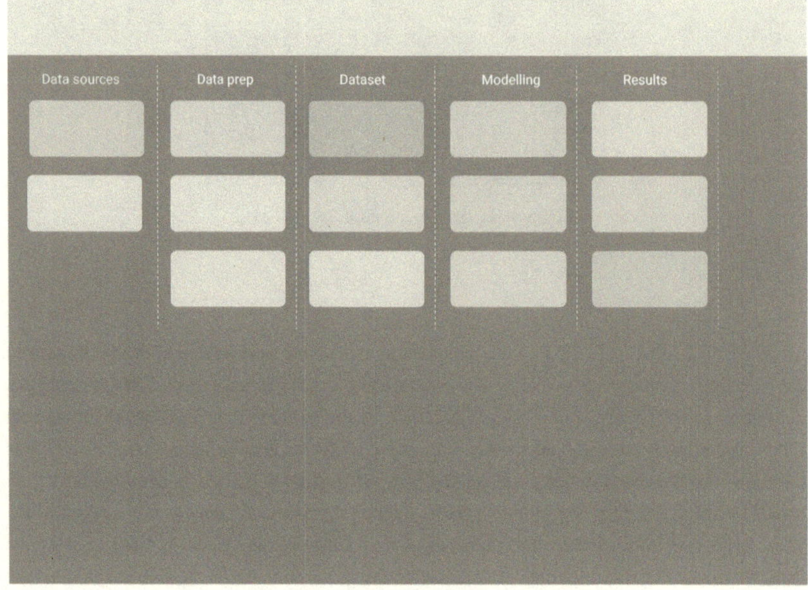

Figure 4.22: COLLAPSE board - normal state

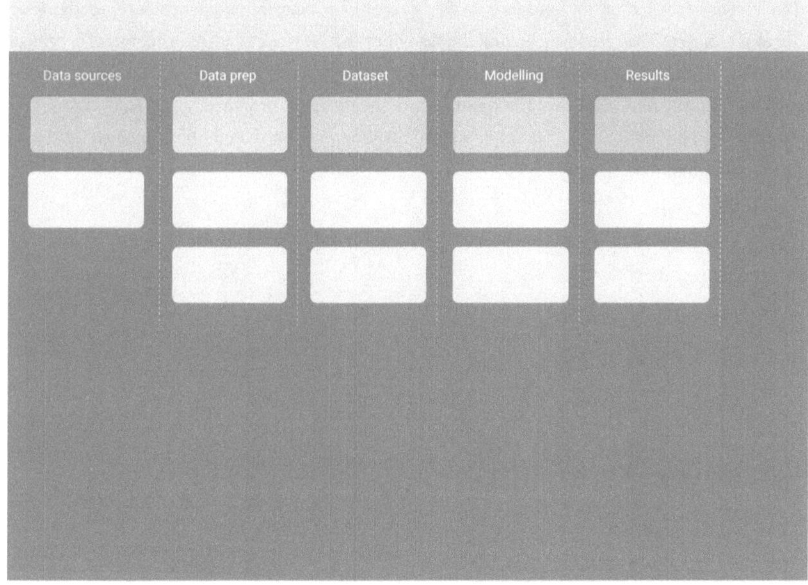

Figure 4.23: COLLAPSE board - after COLLAPSE effect

4.4 Conclusion

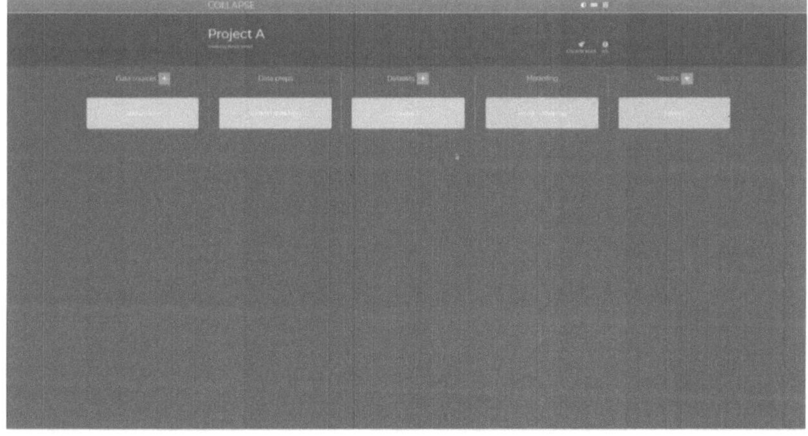

Figure 4.24: Final COLLAPSE board design

This chapter focused on the interface's design, evolution, and the implementation of the functional prototype. The reasoning behind the design of the central hub of information of a project, the COLLAPSE board, was explained. It was shown how the design for the COLLAPSE board evolved from the first sketch seen in figure 4.2, to the mockups designed in Figma (figure 4.1) to the final design (figure 4.24). The implemented features were presented following a workflow for better understanding.

Chapter 5

Tests and Results

This chapter describes the two types tests conducted to validate the platform and the most interesting results extracted from these tests.

5.1 Experimental conditions

To validate the COLLAPSE platform, two types of tests were conducted: usability tests and field trials.

5.1.1 Usability tests

Usability testing is a very popular User Experience (UX) research methodology [21] that allows researchers to uncover problems and opportunities in design. In a usability test, the researcher (also called moderator) asks the participant to execute a series of tasks using a user interface.

These tasks represent realistic activities that the user might perform in real life. The tasks defined for the COLLAPSE usability test are presented in table 5.1.

Table 5.1: Usability tests tasks

Task number	Task
Task 1	Determine who was responsible for "Data source 1".
Task 2	You want to examine the data source created by Data scientist 1 by downloading it to your device. How can you do this?
Task 3	You decide to start from the data source already created by Data scientist 1. After applying the data preparation procedure that seems more appropriate to you, you want to create a dataset (by uploading a file) that uses the data source already created and your data preparation which is stored in a link. How can you achieve this?
Task 4	After creating a dataset and using it to train and develop the model, you observe that your results are different from Data scientist 1's, so you're asked to update the board with your results by referring to a link.
Task 5	After further analysis of your results, you discover that something must not be right with the data preparation procedure you used, so you want to update it, as well as update the generated dataset.
Task 6	You find a new data source that might lead to better results, so you decide to share it with the team on the platform by creating a new resource. The source of this data is a database.
Task 7	You combined several sources to update the dataset. How can you do this in the platform?

It is also important to make sure that the participants represent a realistic user. In the case of the COLLAPSE usability test, this means that the participants should have a similar background to the target user group.

Besides guiding the participant through the test process, the moderator is responsible for annotating any feedback given by the participant.

5.1.2 Field trials

A field trial differs from a usability test because it is conducted for a more extended period, the participants are introduced to the platform they will be testing, and it has different objectives. This type of test is more focused on extracting knowledge from a more intensive use of the platform, rather than assessing if the interface is intuitive. The field trial for COLLAPSE had a duration of three days and had the participation of two data scientists.

5.1.3 Participants

Usability tests

The process of validating the functional prototype started with the selection of subjects for the usability tests. Since the platform is aimed at professionals working in DS, the subjects should be familiar with concepts related to this area. In order to fulfill these requirements, five students developing their book with themes associated with DS were chosen.

Table 5.2: Usability tests participants

Participant	Age	Area of study	Level of experience
P1	23	ML / Computer vision	Medium
P2	22	Natural language processing	Medium
P3	23	ML / Explainable AI	Medium
P4	21	ML / Reinforcement Learning	Advanced
P5	22	Data mining	Medium

Field Trials

Two data scientists working on two different projects were designated to participate in the field trial.

5.1.4 Objectives

Usability tests

The main goal of the usability tests was to detect potential usability issues regarding the workflow, the comprehensibility of the different actions and concepts, the adequacy of the information architecture (e.g., the relation between Data Sources, Data Preps, and Datasets) and how intuitive the interface is, since there was no introduction to the available features, only to the concept behind the platform. Additionally, these tests aimed to evaluate if the platform could improve certain aspects of development like collaboration. The usability test consisted of two sets of tasks. In the first set (Task 1 and Task 2), the objective is to evaluate how well the participant can identify and understand the work done by other team members. On the other hand, the second set (Task 3, 4, 5, 6 and 7) aims to assess how well the participant can use the platform's workflow.

Field trials

The main goal of the field trials was to assess if the usability experience would change after being introduced to the implemented features and after using the platform more intensively. This means that no specific tasks were given to the participants.

5.2 Results

The usability of the platform was measured using two metrics: effectiveness and satisfaction.

5.2.1 Effectiveness

Effectiveness relates the platform's goals to the accuracy and completeness with which these goals can be achieved. Since there were no tasks given to the participants of the field trial, the assessment of effectiveness is not applicable to this type of test. Therefore, the following metrics were collected (table 5.3) when conducting the usability tests:

- Completion rate: The proposed task was considered completed when the user successfully completed the task without an error or assist.
- Errors: An error was counted every time the participant performed a task that brought no value to the completion of the task.
- Assists: An assist was counted every time the participant requested the assistance of the facilitator. An assist was not considered if it happened because the task was not well described.

Table 5.3: Effectiveness metrics

Participant	Completion rate	Errors	Assists
P1	57%	5	3
P2	28%	6	4
P3	57%	4	1
P4	71%	3	0
P5	71%	2	3

Given the small pool of participants and the variety in performance they presented, it becomes hard to find reliable patterns that allow knowledge extraction.

However, by analysing the number of errors (figure 5.1) and assists (figure 5.2) per task it is possible to observe that Task 3 and Task 5 present a number of errors and assists much higher than the other tasks. This can represent an aspect that needs improvement. This is later discussed when describing the comments and suggestions made by the participants.

5.2.2 Satisfaction

Satisfaction describes a user's subjective response when using the platform. Satisfaction was measured through the System Usability Scale (SUS) questionnaire administered after the test. Besides the SUS, two extra questions were asked to specifically assess if the platform would improve collaboration and provide a good overview of the project:

- Question 1: On a scale of 1-5 how much do you think COLLAPSE improves collaboration between team members?
- Question 2: On a scale of 1-5 how well do you think COLLAPSE offers an overview of the project?

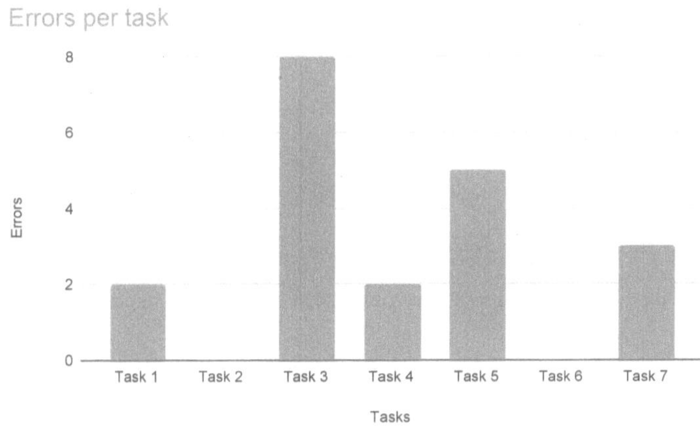

Figure 5.1: Errors per task

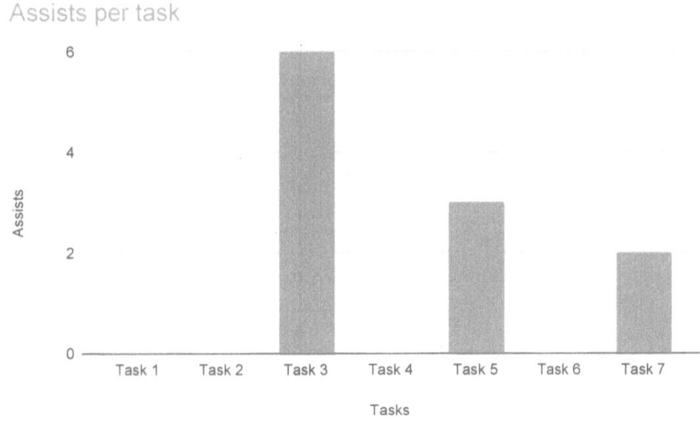

Figure 5.2: Assists per task

The objective of these questions was to validate the objectives set for the platform.

System Usability Scale

Created by John Brooke in 1986 [4], the SUS is a ten item questionnaire (table 5.4) with five response options, ranging from 1 (strongly disagree) to 5 (strongly agree). From these ten items, the odd-numbered items are worded positively and the even-numbered items are worded negatively.

The SUS score is determined in a specific way [4]:

- Odd items: subtract one from the user response
- Even items: subtract the user response from 5
- Add the converted responses for each user and multiply the total by 2.5

Table 5.4: SUS questionnaire items

Item number	SUS item
1	I think that I would like to use this system frequently.
2	I found the system unnecessarily complex.
3	I thought the system was easy to use.
4	I think that I would need the support of a technical person to be able to use this system.
5	I found the various functions in this system were well integrated.
6	I thought there was too much inconsistency in this system.
7	I would imagine that most people would learn to use this system very quickly.
8	I found the system very cumbersome to use.
9	I felt very confident using the system.
10	I needed to learn a lot of things before I could get going with this system.

Usability tests

The SUS scores obtained in the usability tests were computed and are presented in table 5.5.

Table 5.5: Usability tests SUS scores by participant

Participant	SUS score
P1	75
P2	85
P3	75
P4	87.5
P5	82.5
Average	81

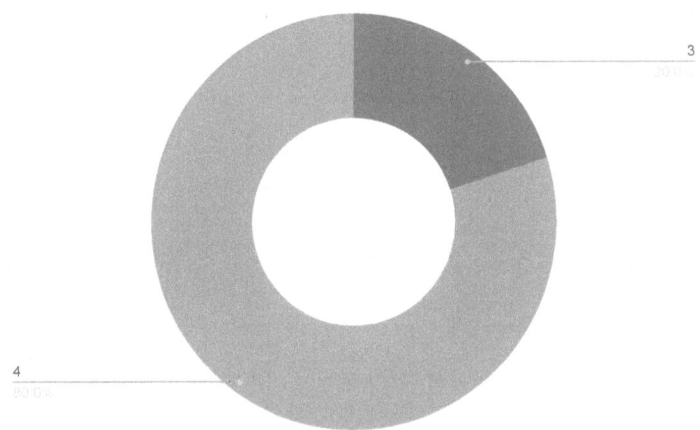

Figure 5.3: Question 1 answers from usability tests

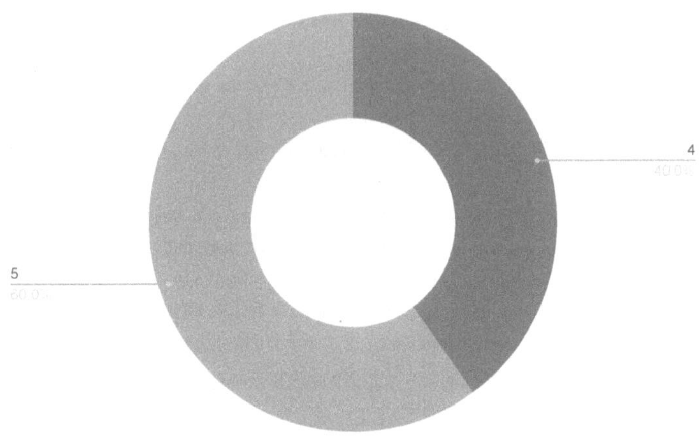

Figure 5.4: Question 2 answers from usability tests

Field trials

The SUS scores obtained in the field trials were computed according to the method previously mentioned and are presented in table 5.6.

Table 5.6: Field trial SUS scores by participant

Participant	**SUS score**
P1	92.5
P2	92.5
Average	92.5

Table 5.7: Answers given by field trial participants to Question 1 and 2

Participant	**Answer to Question 1**	**Answer to Question 2**
P1	4	4
P2	4	4

The average of all the user's scores represents the score for the platform. This means that the COLLAPSE platform has a SUS score of 81, well above the average for this kind of test, which is 68 [21].

The best way to look at a SUS score is to convert it into a percentile through normalization. This means that a SUS score can be converted into a letter classification from A+ to F, where 68, the average value, represents a C. To get an A the score must be above 80.3, therefore COLLAPSE gets an A in usability, since its score was 81.

By analyzing figure 5.3 it is possible to see that 80% of the usability tests participants think that the platform promotes collaboration between team members at a pretty high level, as they answered a 4/5. Both of the field trial participants agree with this, giving the same answer, presented in table 5.7.

By analyzing figure 5.4 it is possible to see that 60% of the usability tests participants think that the platform provides an excellent overview of the project a 5/5. Both of the field trial participants partially agree with this, answering Question 2 with a 4/5 (table 5.7).

All the comments made by the participants during the test and the informal conversation that followed were registered. In these comments, the platform's intuitiveness was mentioned by more than one participant, also being said that if given enough time to mature in the real world, the platform could evolve in a positive way.

As aspects to improve are concerned, several participants denoted that when creating a Dataset (Task 3) or Results, it is unclear at first that the form follows the same column organization as the board. For example, there were some problems identifying the destination for the data sources. In addition, some participants pointed out that, for example, having to edit a dataset (Task 5) to be able to edit the data prep used in that dataset is odd when not familiar with the workflow.

Despite giving this feedback, several participants stated that a small workshop introducing the platform to the team could solve some of these issues. This is backed up by the SUS scores of the field trial participants, that scored higher than the participants of the usability tests.

Chapter 6

Conclusions and Future Work

Data science projects apply the scientific method, and this focus on experimentation is what differs them from a typical software development project. The technological and conceptual differences make it so these projects have different needs to fulfill. Several methodologies were defined and evolved throughout the years in order to keep up with the evolving needs of these projects. Tools were also created to help data science teams have a smoother development process. Even though these tools bring a tremendous amount of value with their features, they lack value when it comes to project management. Therefore, the main objectives of this project were to develop a tool that would promote collaboration in data science teams that could provide an easy-to-understand overview of the project while being flexible, allowing each data scientist to work in their preferred environment. Given the temporal limitations, not all the features could be implemented, so a set of these features was chosen to be implemented, building a functional prototype that represents the Minimum Viable Product (MVP). Despite this, the development process and architecture design were defined considering all the functionalities of the platform.

The analysis of the methodologies used in DS projects lead to an understanding of their evolution and how the platform should support them. Additionally, the analysis of the most common tools and platforms used by teams revealed itself very valuable, since it allowed for the definition of the place of COLLAPSE in the DS ecosystem.

Since the platform was meant to be collaborative, it became clear that COLLAPSE should be a web platform. This would not only facilitate its implementation, but its testing as well. The user stories representing the platform's features were defined after the definition of the idea.

The architectural pattern defined for the platform was a three-layered architecture to facilitate the future development of new features. These three layers consist of a presentation layer, an application layer, and a persistence layer. The presentation layer was developed using React, the application layer was built with Express, and the persistence layer uses PostgreSQL.

Finally, despite having a small sample of participants, the usability tests and field trials revealed that the platform could bring value to the DS ecosystem.

6.1 Future work

Given the reusability and expandability of the developed frontend components along with the chosen platform architecture style, it is safe to say that this project represents a valuable base to build upon. Therefore, future work can be divided into two types: short-term and long-term work.

In terms of short-term development, the remaining features should be implemented. This means that user stories US3, US24 and US25 should be tackled in order to implement editing a project (US3), adding collaborators to the project directly on the platform (US24) and the COLLAPSE effect previously discussed (US25).

Besides this, the issues extracted from the usability tests feedback should be addressed.

When it comes to long-term development, the goals should be more ambitious, like allowing the user to download a resource in several file types and integrating with any of the tools mentioned in section 2.2 to bring even more value to COLLAPSE and the DS ecosystem.

www.ingramcontent.com/pod-product-compliance
Lightning Source LLC
LaVergne TN
LVHW041252150826
845673LV00008B/2556

* 9 7 8 3 3 8 4 2 0 9 2 5 2 *